Willing Partners: Insight on Stockmanship

May 15/2013

To Julie
Life's a Journey

Peter Campbell Talks Horses

Willing Partners: Insight on Stockmanship

Published by

The Frontier Project Inc.

Inquiries: The Frontier Project Inc.
PO Box 7160, Pueblo West CO 81007

Designed by Emily Kitching & Steve Bell,
Eclectic Horseman Communications, Inc.
www.eclectic-horseman.com

ISBN: 978-0-9853425-2-4

Peter Campbell and "Dillion," Hat Creek Ranch, Wheatland, Wyoming.
Photo by Trina Campbell.

Contents

Peter Campbell aboard "Fancy," Calgary Stampede Ranch, near Calgary, Alberta, Canada.
Photo by Trina Campbell.

Dedication

I want to dedicate this book to all the horses that have truly been my best friends and my greatest teachers. A few of them you might know: Candy, Fancy, Dillion, Molly, Clara, Jackie, Superman, Snorty, Hercules, Doc, Magic, Robbie, Dixie, Anita, Betty, Frankie, Brownie, Dolly, Cindy, Henry and Tom.

Foreword

As a saddlemaker, I make many friendships through my work. So many of my acquaintances have started with saddle orders and, as the years have gone by, many customers have become very special friends.

I first became aware of Peter Campbell back in the early 1990s, when he ordered two saddles from me. Over the course of building his saddles, I spoke to him a few times on the phone, but never met him in person until the first saddle was finished and he and his wife, Trina, came to pick it up.

My first impression of Peter is the same impression I have to this day: He's a motivated horseman, honest and sincere.

Over years of watching Peter work, I've come to believe that his role as a "horse clinician" is misunderstood. He's not there just to help the horse. He does clinics to help individuals, his students, so they can help their horses.

Few riders understand the importance of the term "getting down to the feet," a phrase that was used by both Tom Dorrance and Ray Hunt. Until a rider can tell where each of a horse's feet is, and can become able to position the horse's feet, that horse will always present some resistance. Peter is exceptionally gifted at demonstrating to his students where a horse's feet must be in order for them to make specific turns or maneuvers. It's fundamental, but so difficult to achieve.

Dale Harwood

Shelley, Idaho

Peter Campbell and saddlemaker Dale Harwood, Hat Creek Ranch, Wheatland, Wyoming. Photo by Trina Campbell.

There are a million different ways to work a horse. For me, there's only one right way: to work from where the horse is at.

- Peter Campbell

Peter Campbell, Hat Creek Ranch, Wheatland, Wyoming. Photo by A.J. Mangum.

Chapter One

The Beginning

I ***was born in Banff, Alberta,*** a resort community in the Rocky Mountains, west of Calgary. Our neighbor ran a rental car franchise at the Banff Springs Hotel and, when I was about nine, he gave me a job selling parking-lot tickets to tourists. Near the lot, there was a horse corral.

I remember there were no horses in the corral for most of my first day on the job. I spent the day selling my tickets and, eventually, up a nearby trail came three strings of horses, tied head to tail, with 10 head in each string. I was intrigued, and from my perfect vantage point in the parking lot, I watched as riders brought the strings into the log horse corral. Something about that moment set in motion a chain of events that's defined my working life.

In addition to running pack strings, the outfitter drove teams. He and his crew would pick up tourists and take them to what was called the "doughnut tent," a canvas tent with a hole in the center, through which smoke from a campfire could escape. The tent was the venue for western-themed parties complete with bands. Once school started again, I'd ride along on the wagons. Soon, I was learning to drive the teams.

My work in Banff evolved over time. I began spending summers helping around the horse corrals. The operation was sizeable, with 300 or more horses. I learned about pack horses and backcountry riding, and spent more time driving teams.

When I was 16, a I hired on with the park service. The job was straightforward, but unique. It involved riding in the backcountry, clearing deadfall from trails. I'd spend nights in warden cabins, which were scattered throughout the backcountry. Each had a corral, a tack shed and a little pasture. In addition to being an incredible adventure for a teenager, it gave me a

Peter Campbell, Banff National Park, Banff, Alberta, Canada. Photo from Audrey Campbell family collection.

rare chance to see the entirety of Banff National Park, and to ride all its trails.

The park service horses came from a government-run Alberta horse ranch called the Ya-Ha-Tinda, located on the eastern slope of the Rockies. The ranch ran a stallion and broodmares, trained horses, and wintered the veteran park horses. The ranch intrigued me and, when I was 18, I landed a job there working horses. To this date, I might be the youngest ever to be hired as a horseman at the Ya-Ha-Tinda. Looking back, I'm lucky I'm not dead. I'd ridden lots of miles and plenty of horses for my age but, relatively speaking, I knew little.

The job entailed looking after the mares and foals. And, each of us on the ranch had seven or eight colts under our care. We spent our days riding those colts on the ranch's 18,000 acres of rolling hills and grass pastures.

In the winter, the saddle and pack horses from the parks arrived at the ranch. We looked after them until spring, when

Peter Campbell at Deception Pass, near Banff, Alberta, Canada. Photo by Brad White.

we sent them back into service, along with the young horses we'd ridden the previous year. With that bunch gone, we'd start the cycle again, riding a new set of colts for the next year. That was the real start of my professional life as a horseman.

Working with the horses at the ranch, I knew enough to realize that if a horse got scared or ran off or went to bucking, I needed to bend the horse and get him stopped. I don't know how I knew to do that, but I knew that when a horse was about to get himself in trouble, I needed to change his mind.

Occasionally, though, there were problems with horses, problems that I couldn't seem to solve, no matter what I tried. I asked a friend for advice and he told me he'd heard of a guy -- a talented horseman -- that was coming to Alberta to put on a clinic. He was someone who might be able to help me better understand how to work with troubled horses. The horseman was Ray Hunt.

Everything I saw at Ray's clinic made sense to me, but reproducing it was another thing altogether. Once Ray had

Peter Campbell (right), Ya Ha Tinda Ranch, Sundre, Alberta, Canada.
Photo by Bruce Campbell.

worked with a horse, that horse was quieter, smoother, less troubled. If a horse was quiet to begin with, most people might not notice much of a difference. But if you began with one that was troubled, the difference was huge: the horses worked smoother and seemed more inclined to want to get along with people. They were less inclined to be bothered or upset. I could see Ray's approach helped those horses, and I couldn't wait to get back to the ranch to work with my horses in a whole new way.

The more I tried to use the techniques I'd seen at Ray's clinic, the more intrigued I became. I was eager to learn more, so I traveled south to California to meet Ray's mentor, iconic horseman Tom Dorrance.

The first day I met Tom, he was working with a horse that was "bracey" through his loin and hindquarters. A man held the horse by a long lead rope as Tom instructed him on how to maneuver the horse. Tom noticed the horse was beginning to free up and told the handler to let the horse "drift." The horse

Ray Hunt (left) and Peter Campbell, Nanton, Alberta, Canada. Photo by Brad White.

rolled over behind, the brace seemingly gone. At the time, I didn't have a clue what Tom was doing. It made no sense whatsoever.

All I knew was that, somehow, Tom could see that the horse was beginning to get the picture, that as he worked with his human handlers, the proceedings were beginning to have some meaning for him.

After stints working at British Columbia's Gang Ranch, one of the largest outfits in North America, and the Wineglass Ranch in Alberta, I moved to California to be close to Tom and to learn from him. While there, I buckarooed on ranches and rode horses for a living while I developed a friendship with Tom. He'd help me with troubled horses; I'd accompany him when he went to work with a horse. I got hooked. I couldn't get enough of the philosophy at work. Soon, I began helping Tom at his clinics; I rode horseback while Tom directed.

I found Tom's clinics exciting. When he was around, things always seemed to go smoothly. A horse might work on the

Peter Campbell, Scotch Camp, Banff, Alberta, Canada. Photo by Brad White.

wrong thing, but it was never chaotic. It was a valuable lesson in what was possible with horses.

When he worked with horses, Tom could do so little and get so much. A lot of folks today think it must have been all sugar cubes and petting. In reality, in order to help a horse change, Tom would present the situation to the horse in such a way that the horse would feel as if he were putting pressure on himself to adjust his way of thinking. There was no punishment, which occurs after the fact; instead, there was discipline, which happens before or during an act. There was no correction, only direction. It was intriguing.

The thing about Tom was that he could achieve these changes with a horse by getting the idea across to just about any rider. He'd get his ideas through to a person, changing the way he presented them to fit that person's ability. That's a special talent. And, as good as he was with horses -- and cattle, for that matter -- for Tom, it was really about life. His mindset was the same, regardless of whether or not a particular situa-

Tom Dorrance (left) and Peter Campbell, 4W Ranch, Sundre, Alberta, Canada.
Photo by Bruce Campbell.

tion involved horses.

I recall driving somewhere with Tom. I was behind the wheel and we were stopped at an intersection. I wanted to get going and make my turn. I sat there, saying, "Come on, come on," and tapping the steering wheel. Tom turned to me and said, "Just fix it up and wait, Peter. When the coast is clear, we'll go." It was the same patience he displayed with horses, the same confidence that a solution is coming -- you just have to wait for it and know it when you see it.

As time went on, I began putting on clinics of my own when I wasn't buckarooing on California ranches. In 1995, clinics became my full-time profession.

A clinician begins his career with the idea that his students crave knowledge as much as he does. That leads you to push your students, and it can come across like you're picking on them. Really, you're just trying to get them to offer what their horses need so badly from them.

You finally realize that the more pressure you put on

Peter Campbell (left) and Ray Hunt, Cochrane, Alberta, Canada. Photo by Judy Gunn.

students, the harder it is for them. So you ease off and begin reveling in the smallest of accomplishments -- helping someone move from the hackamore to the two-rein or even helping someone learn to tie a halter properly. It's just like with horses: reward the smallest change, the slightest try.

When I first started, there were a lot of bothered horses at clinics. Times are changing, and I don't see as many of them as I used to. People are getting educated. And, they're learning to focus their efforts, staying on one path rather than trying to adopt multiple approaches from multiple sources. I tell people that it doesn't matter what path they pick, as long as they pick one. Otherwise, it's like trying to drink a punch that has a little bit of every kind of alcohol. In short order, it's not worth drinking.

The people that come to my clinics are really there for their horses, to see what they can do for their horses, to help them stay right. And, to me, that's what it's all about.

My experiences with students have made me understand,

Peter Campbell and "Dixie," 5 Dot Land & Cattle Co., Benecia, California.
Photo by Trina Campbell.

though, that we're still just scratching the surface of what my friend Tom knew about horses. Tom passed away in 2003 and no one on his level is coming down the pike. I'm not Tom, and all I can offer is my interpretation of his work. I do know that, if we're not careful, if we're not as true as we can be to what he helped many of us understand, the wisdom he shared could fade from the landscape. And that would be an immeasurable loss for both riders and horses.

Peter Campbell and "Molly," Hat Creek Ranch, Wheatland, Wyoming.
Photo by Trina Campbell.

Whatever the horse is offering us,

it's bigger than us.

- Peter Campbell

Peter Campbell, Hat Creek Ranch, Wheatland, Wyoming. Photo by A.J. Mangum.

Chapter Two

Horsemanship

To me, "horsemanship" means the development of one's self. It's a simple idea. We're working on ourselves, developing the discipline that allows us to understand what the horse needs from us and how best to present ideas to the horse. Achieve that, and you'll find that working with your horse becomes a whole lot easier.

In my earliest days working with horses, in the 1970s, Tom Dorrance's methods weren't yet prevalent. His methodology was still under the radar in much of the horse world. Horse handlers tended to try and dominate horses, and used some approaches that were pretty harsh. Even then, I never liked trouble with horses, never liked to see one that thought he needed to buck or feel like he was in a bad situation. Finding a path that was clear of those troubles, though, was a challenge.

When I was around 14, an outfitter I worked with had a leased pack horse named Lucky. Everyone in the outfitting operation seemed to have trouble getting along with Lucky, and considered him nearly impossible to shoe. The horse, though, offered a different experience to his owner, an older fellow. He could get the horse shod, where few others would even make the attempt. I remember thinking that, where Lucky was concerned, there was obviously something everyone but the old man was missing.

In those days, the preferred solution for most "problem horses," was to get on and ride, to put on miles and put in time. The reality, though, is that miles and wet saddle blankets aren't what make good horses. If you're working at the wrong thing through all those hours and miles, a horse will simply get tougher and tighter. If you work at the right thing, though, and a horse gets some meaning out of it, the lessons will stick. It won't be a function of hours and miles.

Over time, you start to see what you need to offer the horse to help him do what you're asking of him. You learn to see where the horse is coming from, and learn to recognize the ways his instinct for self-preservation shows itself. It all begins to make sense. When your horse encounters something that makes him uncomfortable and he responds positively to your encouragement, giving you his trust, you're in a delicate posi-

Peter Campbell and "Magic," Ellensburg, Washington. Photo by Trina Campbell.

tion. If you get your horse into trouble after he's given you that trust, he begins to question your judgment. After all, the last time he trusted you, it got him into trouble. What riders often define as problem behavior in a horse is really the horse's attempt to protect himself.

Working with horses presents a unique challenge. When you begin, you really don't know what you don't know. Ironically, it's by making mistakes and exercising bad judgment that you begin to collect the experiences that make good judgment possible. Those lessons build upon one another, and it often takes later experiences to shine light on previous lessons and give them meaning.

Years before I knew Tom, I had a little bay mare that had a habit of rearing. It wasn't violent, but she'd jump in the air. I took her to a friend for help. Standard procedure then was to over-and-under a rearing horse. He did, and she stopped, but the problem wasn't solved. On later rides, she still reared with me.

When I worked with Tom, I rode a horse that had the

Peter Campbell and "Jackie," Ann Arbor, Michigan. Photo by Meghan Ward.

same problem with rearing. I shared with Tom my previous experience with the bay mare. He explained that over-and-undering a horse might get it to move, but when you really needed to count on the horse, that kind of negative behavior, rearing, would always surface again. Over-and-undering wouldn't have taught the horse not to rear, Tom explained. It would've only made him afraid, and would've done nothing to change the horse's mind about rearing. Without having had that previous experience with the bay mare, and without having made mistakes in trying to manage her rearing, I might not have understood what Tom was saying to me.

Working with horses has taught me to have discipline in everything I do; to be particular, but not critical; to revel in small, rather than big accomplishments; to keep offering myself to the horse and, even though the horse might not be at a hundred percent, to keep offering; and to not force things. It'll always take time to get somewhere but, one day, you'll get on your horse and you'll feel a difference. The horse will have turned loose, mentally. Patience might not be the word. You

Peter Campbell and "Candy," Fort Collins, Colorado. Photo by Trina Campbell.

do end up waiting on your horse, but waiting isn't necessarily doing nothing. You're just not making things happen.

Unfortunately, when a horse resists, most people will do one of two things: quit, or go to beating on the horse. If you can instead simply wait, your horse will make it. And it's when the horse is in the most trouble, when he's in his darkest hour and putting up the most resistance, that he'll often make a breakthrough. It's in those moments that you need to be at peace, to keep offering, and to reward every little effort. It's those little steps, those little victories, that help a horse understand what you're trying to get across to him.

It can be the same with people. Early in my attempts at teaching, I tried to make riders understand. I figured that, since I had this driving fire to figure this out, everyone else must have the same desire. When a student had a problem, I'd push hard to help them.

Now, I'm able to let students work at trying to understand and, as long as it's not a dangerous situation, I can let them find the answers and revel in the small accomplishments they

Peter Campbell and "Candy," Hat Creek Ranch, Wheatland, Wyoming.
Photo by Trina Campbell.

make along the way. If they make 30 seconds of progress, I celebrate that, rather than obsessing over making all the progress possible, all at once.

At a clinic, a student might be working with a horse, trying to make something happen. I'll tell them to keep offering. The horse might not make his breakthrough at the clinic, but I might come back a year later and see that things are working. The horse is now walking in a straight line or stopping smoothly. And from the look on the student's face, I can tell that his insides have changed. He's no longer frustrated.

Some horsemen reach a point where they feel like nothing can stump them. They figure they've been around, seen it all, and that there's nothing they can't figure out. I've learned, though, that as you spend more time with horses, you realize there's always so much more to know. Whatever ego you begin with, horses will eventually get it in check and you'll leave every situation feeling like there's still so much to learn.

Peter Campbell, Hat Creek Ranch, Wheatland, Wyoming. Photo by A.J. Mangum.

When I ride my horse, I don't get on him to see what he can do for me. I get on him to see what I can do for him.

- Peter Campbell

Peter Campbell at work, Hat Creek Ranch, Wheatland, Wyòming. Photo by A.J. Mangum.

Chapter Three

Colt Starting

W**hen I began conducting clinics,** you'd see some students bring older, unhandled horses to be started under saddle. That's changed. People are better educated and most of the horses I help start at clinics are two, maybe just a little older.

More important than a horse's specific age, the key thing is being sure you've prepared the horse. He needs to be halter-broke so you can use the halter and lead rope to move his feet, to move his hindquarters and front end, and get him to shift his weight, take a step back, and lead forward. Get the horse to where you can touch him with a flag or the coils of a rope and get him prepared for the saddle.

Of course, you can start handling a colt when he's a week or two old. You don't want to force your will upon one, but if you don't make a point of trying to touch them, the next thing you know, they get curious, want some attention, and come up to see you.

At home, we halter break colts as yearlings, play with them a little, and then, when they're two, we'll play with them again, rubbing them, touching them, preparing them for the saddle. That's the toughest part, because you don't want a horse to be afraid. If he's scared, you can't get him to shape his body. But if you expose him to those ideas in a manner he can understand -- for instance, using the flag or the coils of your rope -- he can still have a sense of self-preservation as he finds out these things aren't going to hurt him. A horse wasn't born with a saddle, but he needs to get used to the idea of its presence, just like he's used to his own mane and tail. Accomplish that preparation, and saddling itself doesn't have to be a big deal. Once you've got a horse ready, the saddle can just go on.

Of course, you don't want to start a horse that's sore or lame or in poor condition. Colts need to feel their best when you go to start them. They're like us. If you missed dinner for three days, you wouldn't feel up to snuff. If someone has a colt that's a little down or sick or not as full as he should be, that person needs to wait on him.

Some horses might be too small to ride, but their handlers can still work with them, get them used to being handled, get them used to a flag or a rope, get them used to carrying a snaffle bit in their mouths.

Get the horse to want to be with you, whether you're on the ground or on his back. I try to get him to understand that

Peter Campbell starting a colt while aboard "Superman," Kelowna, British Columbia, Canada. Photo by Mark R. Lawrence.

he doesn't need to be afraid of the things I'm asking him to do. If he does get afraid, I do a little less and that fear will turn to respect.

I use things he might be a bit bothered about to get control of his feet. If he's worried about the flag, for instance, I'll use that. I don't try to scare or upset the horse, but he might get scared or upset. There's a difference. When I get control of his feet, by the time I have his feet "hooked to" the halter, he's usually not scared of that object anymore.

You can use your rope, touching him with the coils, to check him out and get him ready to saddle. You can use your flag. You might use your halter rope around his belly to check him out, and see how he might react to being cinched up.

At clinics, I tell students that if they get too far back while they're working on the ground, they can get kicked. If they get too far ahead, a horse might strike them or bite them. There's a fine line. If a horse is touchy, they can use the fence -- get the horse up to the fence to touch him with the flag to get control of his feet and hindquarters.

That's why we use tools like a flag. It makes your arm a little longer. If a horse is going to get scared, it gives you a

Peter Campbell aboard "Charlie" at a clinic in Ann Arbor, Michigan.
Photo by Diane Thomson.

little more clearance. You have a little more room to make a recovery if you're in the wrong spot or if a horse gets scared.

As for round pens, I don't think it matters what shape the pen is. It's more about being aware of where you need to be. The round corral can make it easier for a handler since the horse can't get caught up in a corner. If it's a square pen and someone's in the wrong spot, a handler could push the horse up over the fence. But that can actually happen in a round pen, too.

The way some people use a round pen, sometimes the horse is just in automatic mode. He's been around the pen so much that he shuts off mentally. He's just going round and round and thinks that's what it's all about. To me it would be like working in a factory, putting the same part in the same spot day after day. Before long, you don't think about it much. With horses you have to be careful not to overdo it.

When the horse is waiting there, put your saddle pad on, put your saddle on, and cinch him up. Most of the time, a horse is good to stand there and be saddled. They don't know what's happening, they've never had it happen before and, often, don't see a reason to be scared, as long as you're smooth

Peter Campbell works with a colt, Hat Creek Ranch, Wheatland, Wyoming.
Photo by Trina Campbell.

when you put the saddle on. If you've done your homework and checked him out -- and, again, if you're smooth -- the most common scenario is that he'll just stand still.

When a saddled colt first moves, that's when he can get bothered or upset. You have to adjust to fit the situation. If it's a nice, safe pen, you might cinch him up and move him a couple of steps on the end of a lead rope.

Or you might turn him loose and just let him carry the saddle as you move him around a bit. He'll explore a little and find his way of moving his feet.

You can also turn him out in a bigger pen and let him move around with the saddle on his back. In any case, it's like he has to get used to a new part of his body. The saddle's doing things his mane and tail have never done.

Next, you can catch the colt and work with him a bit to get control of his hindquarters and front end. Get him so he'll walk around you in a circle and step back a little bit. Now you can start preparing him to have someone get on him.

Lean against his side, rub his neck and rub his hindquarters. Get him used to being touched there because when you get on, your chaps might brush him across his hindquarters.

Peter Campbell aboard a colt, High Country Quarter Horse Ranch, Eastend, Saskatchewan, Canada. Photo by Leann Gilbert.

Move the stirrup forward and back because your leg will pick up that stirrup when you get on and you don't want him getting scared and leaving the country.

When you go to get on a colt, take hold of the inside rein and keep his nose tipped. But don't bend his head too much or he could get off balance. Put your toe -- not your whole foot -- in the stirrup and lean up against his side. If the colt gets tight, step down. Some folks stay there a little too long and the horse gets upset, like he's about to get broncy. Step down, move his hindquarters over a couple of steps and start again.

As you get on, make sure that when your leg starts over the horse's back you have hold of something solid -- the saddle horn or the fork of the saddle -- so if he gets scared and jumps forward, you won't scoot out the back door. If there's the potential for him to be that scared, though, he's not ready for you to get on.

Remember that no amount of groundwork is going to substitute for you getting on a horse. You don't need to be in a hurry about getting on a colt, but if you're spending months on groundwork, you're turning it into a wearing process rather than a learning process. It's time to get on.

Peter Campbell and "Brownie," High Country Quarter Horse Ranch, Eastend, Saskatchewan, Canada. Photo by Leann Gilbert.

Once you're in the saddle, at first, you just go where the colt goes. You're a passenger. If he gets scared, you can rub him, support him a little bit with the reins. That's something you have to manage carefully. If you offer no support at all, they can get lost and scared. A little support helps them. Too much, and you overconfine them, and that can scare them worse than no support at all. With the right amount of support, you can direct them, get them into a walk and trot, and get the hindquarters to step over.

I'd say you need to walk, trot and lope in an arena before you get outside. Work up to a trot, then make it difficult for him to trot so he'll prepare to roll into a lope. Don't try to keep him in the lope. At first, you can just let him roll back into a trot.

Get these things working in a corral or arena. It won't hurt to take a horse out and lead him along behind another horse, with the saddle on his back and stirrups bouncing. Once you're riding outside, you want to be careful, especially with a horse that's been raised in a stall. There's a lot of new stuff to see outside. With ranch horses, of course, they've seen it all. Safety's number one, though. When you're not sure about

Peter Campbell aboard "Doc" and roping "Iceman," High Country Quarter Horse Ranch, Eastend, Saskatechewan, Canada. Photo by Leann Gilbert.

something, or if you're by yourself in a questionable situation, just wait.

Riders naturally question when it might be appropriate to move forward to something new, when they should begin with the next step. The key is that when the horse is waiting and ready to go on to something new, then you need to move forward. If you don't, he'll regress instead of progress.

If you're flopping the saddle pad on and off, and he's stopped and relaxed and not tight, he's waiting to go on. He's as good as he's going to get with that. If you keep at it, he'll start searching for an answer as to why you're still throwing the saddle pad on him. He might think you want him to move when you really want him to keep standing still. That takes time for a person to understand. When the horse is waiting and ready to go on, the human needs to go on and fill in more blanks for him.

Again, don't feel like you have to get in a hurry, but don't feel like you need a drilling process, where you're wearing him out every day, wearing your idea into him. Make it a learning process, not a wearing process.

You can start a colt in the halter, then go to the snaffle or

Peter Campbell and "Iceman," High Country Quarter Horse Ranch, Eastend, Saskatchewan, Canada. Photo by Leann Gilbert.

hackamore before progressing slowly into the two-rein and then the bridle. But it's not as important to get to that bridle stage as a lot of people think. It doesn't matter what you've got on the horse's head, as long as the reins hook to his feet, which can be the case whether it's a snaffle, hackamore or spade.

You need to be consistent but, once again, avoid drilling. A younger horse, you might only ride for 10 minutes, depending on the horse's size, the shape he's in, the kind of country you're in. If you ride a little bit for a couple of weeks, you're good to go.

If the horse learns in a good frame of mind, he's liable to remember that for a long time. Some horses have better memories. If a horse learns in a bad frame of mind, he'll have a harder time understanding, and it'll take longer to achieve progress with him. Note, though, that even though the horse might get bothered or upset, you're not trying to upset or bother him. A big difference. If you've gone about things in the right way and presented yourself in a manner the horse can understand, you're in good shape to move forward.

Understand direction vs. correction. Understand discipline -- which occurs before or during the action -- and pun-

Peter Campbell aboard "Hercules" and working a colt, Grand Prairie, Alberta, Canada.
Photo by Trina Campbell.

ishment -- which happens after the fact. You need direction, not correction. Discipline, not punishment.

If the rider understands those concepts, it's easier for the horse to understand that there are guidelines to live within. Soon, he wants to be okay. He doesn't want to be upset or bothered. He finds out it's a good thing, what we're offering them, and that there's always a place to go with his feet. There's always somewhere to search in his mind to find what we want. He gets rewarded through timing and feel, and he likes that good feeling. He's not tight or scared, and responds with respect. Once he finds that out, 99.99 percent of the time, he'll take the good you're offering rather than getting troubled.

If a horse had trouble when he was started, he can be tight, defensive, unsure of what's going on. Sometimes they can be pretty good at a slower pace, but fall apart when things get faster. If you want to find the weakest link in your horse, speed him up.

When it comes to working with colts, I'd say misunderstandings are more common than mistakes. People too often look for a quick fix. It takes time to help a horse develop without being in a hurry. If you're on a time schedule, it's

Peter and "Fancy" working with a colt, Neepawa, Manitoba, Canada.
Photo by Trina Campbell.

liable not to work out. And, again, there's no hurry to get to a certain place, but a lot of folks wear their ideas in, over and over. They think they'll get a horse right by getting him tired. Wet saddle blankets don't make good horses if the whole time you're riding him he's in a bad frame of mind. The right kind of wet saddle blankets, when he's mentally right, will help him get better.

You have to think about what you want to happen. If you don't have a clear picture, the chances of it happening are next to nil. It's like the blind leading the blind. See what the horse needs from us, and try not to get fixated on one thing.

Of course, this isn't the only way to work a colt. It's always an individual process. You work with your own capabilities. If the horse has a good mindset and you can ask him to do things and he's happy to do it, you'll be pleasantly surprised at what you can achieve. But you have to know what your abilities are.

Peter Campbell and "Charlie," Ann Arbor, Michigan. Photo by Diane Thomson.

I don't think you correct horses. You don't fix horses. You direct them. The horse takes care of the correcting part. He takes care of the fixing.

- Peter Campbell

"Candy," Hat Creek Ranch, Wheatland, Wyoming. Photo by A.J. Mangum.

Chapter Four

The Snaffle Bit

A***gain, it doesn't matter*** what you ride a horse in. It's the way you present it to him so he can understand the feel. In the snaffle, you can direct rein and tip the horse's nose the way you want it to go. He learns to hold a bit in his mouth and keep his tongue underneath it. When you progress, he's more apt to be able to handle something in his mouth.

Personally, I prefer to use an eggbutt snaffle, which won't pinch a horse's mouth. I use a headstall with buckles on each side, a browband, a throatlatch and, to keep the bit from pulling through the horse's mouth, a chin strap, at least as wide as the bit and adjusted so it isn't too tight. Other preferences include slobber straps that aren't too big and heavy and, of course, a mecate. Riders needn't obsess over the specific style of mecate; any style and size that suits you will do. Crossed split reins are acceptable, too, and might even be preferable in brushy country.

If he's never had one in his mouth before, it's important to let a horse get used to the snaffle bit, ideally by letting him wear a snaffle under his halter while he's still in that halter stage. If you simply stick a snaffle bit in a horse's mouth and go to riding, it'll impede him.

When the snaffle's adjusted, there should be no wrinkles in the horse's mouth, but the bit shouldn't be so loose that it hits the horse's teeth. At first, the horse will play with the snaffle, putting his tongue over and under it. If the bit's adjusted too tight and the horse puts his tongue over it, he might not be able to get his tongue back under it easily. If a horse seems to make a habit of getting his tongue over the snaffle, regardless of its fit, you can feed him while he wears a snaffle; he'll figure out how to get his tongue under the bit in order to eat.

It can be a good thing to work a horse in the snaffle from the ground. You can take him on a corner, step his hindquarters over, ask him to back a bit so he gets accustomed to moving with the bit in his mouth. You're arranging the situation so the horse thinks he's putting that pressure on himself. The end result is that he moves away from his own pressure. You don't

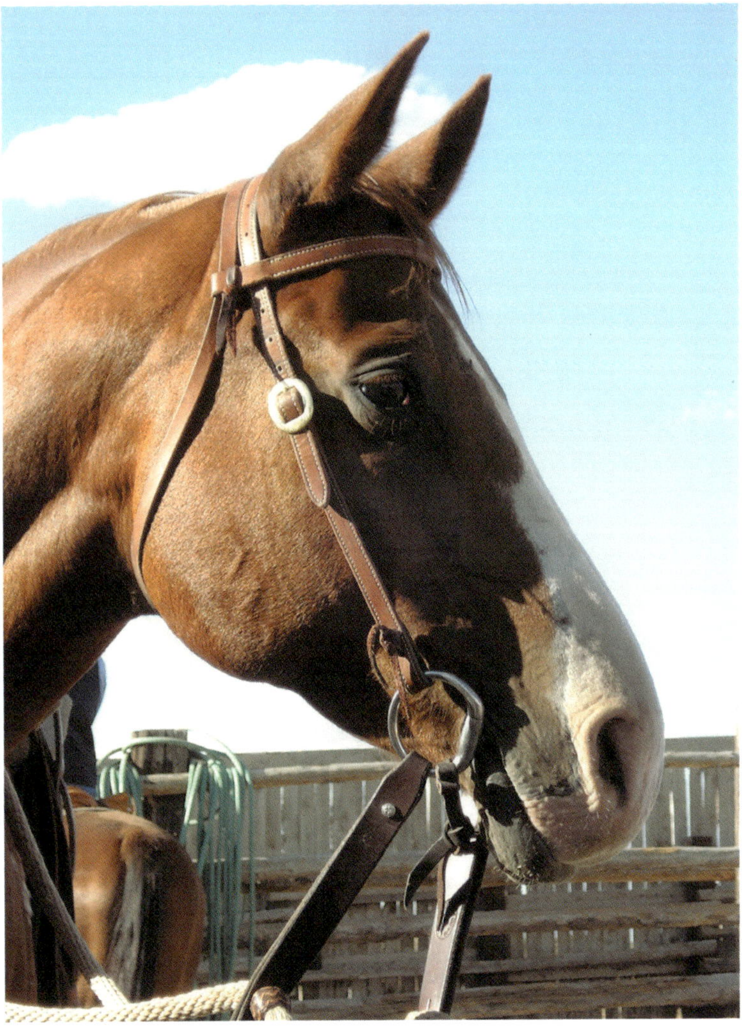

"Fancy," Hat Creek Ranch, Wheatland, Wyoming. Photo by Trina Campbell.

have to drill this into a horse, but make an effort at preparing him for what's to come under saddle.

When you begin to ride with the snaffle in your horse's mouth, the contact with your horse should be feather-like. Give him time to understand the feel of the bit. Don't jerk and snatch. He'll soon put the pieces together. A horse needs time to work at the wrong thing and find that relief that comes through his feet. When he responds on his feet, he gets relief. Don't try to make him respond; let him find what he needs to find.

Many times, riders and horses at the snaffle stage get into tugs of war. The rider pulls on the horse, the horse pulls on the rider. You see the horse getting harder, not more sensitive. As you work, keep in mind that there's no such thing as a hard-mouthed horse. Hard-minded, yes, but not hard-mouthed. Some people decide a horse needs a twisted-wire snaffle to get more sensitivity out of him. And, he might, in fact, be more sensitive for a while but, once his mouth is calloused, he'll get less sensitive, making that twisted-wire snaffle pointless. Again,

"Charlie," Hat Creek Ranch, Wheatland, Wyoming. Photo by A.J. Mangum.

it's not what you put on his head; it's what you put in his mind.

As you ride, always help your horse. You might have to work to get his feet to move first. The life goes from his feet to his leg to his body to his mind. Once that has some meaning for him, the course reverses, going from his mind to his body to his leg to his foot.

With the snaffle, you have a directing rein and a supporting rein. Soon, the directing rein becomes a suggesting rein. The foundation provided by the snaffle's mechanics will serve your horse for the rest of his life. The snaffle can help provide lifelong lessons to the horse: turning, moving his hindquarters, moving his front end, leg yielding, backing, spinning, soft feel, collection. He'll need these skills forever, even if you never advance to the hackamore or bridle.

There's been a push for the last decade or so to make bridle horses, but don't discount the value of the snaffle bit, or the idea of a horse working well in a snaffle for the rest of his life.

With some jobs, the bridle or even the hackamore can

Peter Campbell and "Molly," Loomis Ranch, Wheatland, Wyoming.
Photo by Trina Campbell.

actually get in the way. For instance, when I start colts from the back of another horse, I might use a bridle horse, but put a snaffle in his mouth. I might have to put my bridle horse's head over the colt's back. With a hackamore, that could put pressure on the bosal. With a spade, such contact could also cause unintended pressure on the bridle horse's mouth, and the bit could jab the colt.

If a horse is handled well in the snaffle, a rider wouldn't miss out on anything by keeping that horse in the snaffle. You can ride a horse in a snaffle all his life if you understand that you need to keep the horse sensitive and if the horse doesn't become dull in the snaffle. Can the spade's level of refinement be achieved in a snaffle? Not a hundred percent but, if the rider has handled the horse properly, it would be as close as a human could perceive. It doesn't matter what you have on his head; it's what you've helped him build in his mind that makes him a bridle horse.

Peter and "Anita," Yorkton Grazing Reserve, Yorkton, Saskatchewan, Canada.
Photo by Trina Campbell.

It doesn't matter what path you pick, as long as you pick one.

- Peter Campbell

Peter Campbell and "Dillion," Hat Creek Ranch, Wheatland, Wyoming.
Photo by Trina Campbell.

Chapter Five

The Hackamore

I***f you've started a horse*** in the snaffle and he's working smooth, at a hundred percent, you can put a hackamore on him. Be aware, though, that the hackamore is a different feel for the horse. It's on his nose, with pressure on the outside jaw. In the beginning with a hackamore, some horses have a tendency to tip their heads the wrong way, but with a little time, the horse will start operating like he did in the snaffle and understand that feel.

Some people like to start a horse in the hackamore. What can happen, though, is that the horse ends up running through it. He's not sensitive. He'll push through a turn or stop more than he would with the snaffle and not accomplish either at a hundred percent, unless the rider understands the feel.

Personally, I start some horses in the hackamore and others in the snaffle. If a person starts one in a snaffle and goes to the hackamore, he might think the horse is going to be good in the hackamore, but the horse needs time to separate the difference in feel from the snaffle to the hackamore. And, vice versa: if a horse is started in the hackamore, it will take the horse time to separate the feel of the snaffle from the hackamore. A rider might make that snaffle-to-hackamore transition easier by starting a horse in the snaffle, then moving to the hackamore, then going back and forth a little bit. The horse can separate the difference.

In the last 10 years, the general public has been able to get hold of better and better quality rawhide hackamores. They've always been around, but haven't always been easy to find. You can ride a horse in any kind of hackamore, as long as it's not hurting him, and as long as it's not making his face raw. The hackamore needs to be smooth, but it doesn't need to be expensive. When I move a horse into the hackamore, I usually begin with a 5/8th rawhide hackamore, one that's not so stiff as to wear the hide off a horse's face. Some hackamores, especially those with cable cores, are just too stiff. If your horse operates well on a halter -- which won't be stiff at all -- you shouldn't need a stiff hackamore.

Peter Campbell and "Candy," Diamond 7 Ranch, Cochrane, Alberta, Canada.
Photo by Trina Campbell.

I'd progress from the 5/8ths hackamore down to a 1/2, then maybe a 3/8ths. From there, I'd be using a two-rein hackamore. You want your bosal to fit the horse like a hat should fit your head: it should fit smooth all the way around without putting pressure on any one spot. You should be able to put two fingers between his jaw and the bosal, and the horse should be able to fully work his jaw. If he goes to yawn and relax, but feels confinement, it can scare or upset him.

There's been a certain mystique about the hackamore. It's obviously different in the way it looks and in the way it goes on a horse's head but to me, the hackamore is no different than the snaffle. The reins still "hook to" the feet. It's still a means of communication between the rider and the horse. You just have to be careful when you first use a hackamore so the horse understands that a different feel is at work.

Riding a horse in a hackamore, the rider is liable to find out where any foundation work might've been missed with the horse when he was in the snaffle bit. If he stopped well in the

Peter Campbell aboard "Candy," Marion, Illinois. Photo by Trina Campbell.

snaffle, but it was only 60 percent right, it'll only be 30 percent with the hackamore. You need a horse's foundation at a hundred percent before advancing to the hackamore or those missing pieces will show up and be magnified.

Similarly, before moving forward to the two-rein, a rider wants to be sure of his horse's foundation in the hackamore. Assess a horse's walk, trot and lope; his work at both slow and fast speeds; leg yields; and movement of his hindquarters and front end.

Peter Campbell and "Candy," High Country Quarter Horse Ranch, Eastend, Saskatchewan, Canada. Photo by Trina Campbell.

I'm as true as I can be to what Tom Dorrance helped me understand, keeping my integrity intact and not being taken down another path for marketing, money, fame or fortune.

- Peter Campbell

"Tango," Hat Creek Ranch, Wheatland, Wyoming. Photo by A.J. Mangum.

Chapter Six

The Two-Rein

T***ypically, people have relied*** upon age guidelines to justify moving from one stage of bridle-horse progression to the next. "Between two and three, the horse should be in the snaffle; three and four, the hackamore; four and five, the two-rein; five and six, straight up in the bridle." I take each horse as an individual and work him from where he is at.

When a horse is right in the hackamore, and he's smooth and you feel comfortable performing any task, you can go to the two-rein. Your horse is already used to a smaller hackamore by that point, so let him pack the bridle in his mouth, with its own set of reins. Let him wear that bridle in slow work. As time goes on, you can pick up on the bridle and still have the hackamore to fall back on for faster work, as he's still getting used to the bridle.

Early in the two-rein stage, a horse might be bothered about that bit in his mouth and the rein chains hanging from it. If your horse had issues with a snaffle -- slinging his head, for example -- he'll do the same thing in both the hackamore and the bridle. Again, it comes down to foundation; the reins are not hooked to his feet.

One common problem that occurs: A rider will make this transition from the hackamore to the two-rein, and when things go south and the horse gets bothered, human nature is such that the rider will try to rely on the bit, even though the horse isn't ready for that. The rider's using a two-rein, but if the horse isn't responding to the hackamore, that rider will take hold of the bridle reins. If you're not going to use the two-rein the way it's meant to be used, you might as well not bother with that step.

Some riders find the bridle reins and mecate to be a potentially confusing handful, and it can be cumbersome when a rider has to switch hands for one task or another. Get used to that handful by doing some slow work, and by practicing gathering up your reins, changing hands, and drawing slack. Take

Peter Campbell and "Henry," Erda, Utah. Photo by Trina Campbell.

your time getting familiar with just holding those reins. Find what's comfortable for you.

If people give themselves time to get accustomed to the feel of the reins and the way those reins need to be used, they'll be able to enjoy the two-rein stage. It's a phase of horse-handling that offers a certain sense of pride. You're honoring the horse.

The biggest change you see going forward in this progression is that, in the rider's mind, he's reaching a pinnacle. He's on his way to making a bridle horse, and that's energizing. It can motivate a person to keep reaching for the next stage. It's good, as long as a rider doesn't get caught up in that mindset and channel that energy in the wrong way. You're not in competition with anyone except yourself. We accept that humans are goal-oriented and that this progression from snaffle to bridle offers a goal. But the goal in this case is to get your horse right.

Peter Campbell and "Fancy," near Boulder, Colorado. Photo by Trina Campbell.

The bridle-horse process is about working on yourself, not on the horse.

- Peter Campbell

"Candy," Hat Creek Ranch, Wheatland, Wyoming. Photo by A.J. Mangum.

Chapter Seven

The Spade Bit

The ***spade bit*** can be an intimidating piece of equipment for the inexperienced. Those unfamiliar with it often question how it can be comfortable, or even humane, for a horse.

Twenty years ago, I rode a lot of horses in the two-rein or bridle. People would assume the horses I rode were hard to handle because of the size of the bits in their mouths. They'd see the bit as a leverage tool, whereas it's really a means of offering a suggestion. Make a suggestion to a bridle horse, and he does it, with no force or leverage.

Riders need to understand that the horse's mouth is deep and his tongue is long. A spade doesn't come anywhere close to choking a horse and, in fact, its architecture allows it be comfortably packed by the horse. The horse finds the balance point at which he can comfortably hold the bit. This allows a rider to offer the lightest of signals, which travel from his fingers, down the reins, to the rein chains and to the bit.

When you put your horse in the bridle bit, it's a means of signalling, a means of suggestion. You shouldn't have to take hold of it if he has a full grasp on the feel. You use the reins lightly to offer a suggestion of what to do. Whether it's a half-breed or spade, it's a method of communication.

The motivation behind using the bridle, and developing a horse that can be ridden in the bridle, is to try to get the horse to understand what you want him to do. Getting there is a slow process, but it's an effective means of putting knowledge in a horse without using force.

If someone's not familiar with a bridle's purpose, or its correct use, they can really misuse a spade or half-breed bit. When something's not working in a snaffle, for instance, it seems to be human nature to reach for a bigger bit, thinking that more leverage is surely the answer.

The truth is that if a horse isn't right on his feet, not right mentally, the bit doesn't matter. Reaching for a bigger bit

Peter Campbell and "Candy," Hat Creek Ranch, Wheatland, Wyoming.
Photo by Trina Campbell.

won't fix anything and, in fact, can easily make matters worse. You need to take time to get the horse's feet right, to get his feet to respond to what you're asking him to do, to work without being afraid or numb or dull. If he's dull in the snaffle or hackamore, you can't expect different results in the bridle.

In using a bridle correctly, you're also honoring the horse by creating an opportunity to figure things out without forcing him to do anything. In the mainstream, unfortunately, this is a point that's often lost. Many horses today wear bridle bits, but they're not bridle horses. They're ridden in snaffles, then long-shanked broken bits of one kind or another, then bridle bits. They're not brought up in a traditional way -- riding them in the snaffle, then the hackamore, the two-rein and, finally, the bridle -- and the process tends to be rushed, at the expense of both the horse and the rider's horsemanship.

Horses get more confidence if you take your time and work toward continual improvement, which might or might not involve striving toward riding in a bridle.

Peter Campbell and "Candy," Hat Creek Ranch, Wheatland, Wyoming.
Photo by Trina Campbell.

The bridle-horse process is about working on yourself -- not the horse -- so that you gain the knowledge you need to help your horse understand what you want him to do, without using force and without inducing fear. It's a time-honored concept and if we don't keep up with it, it'll get lost because of the way people misinterpret it.

Bridle horse, Hat Creek Ranch, Wheatland, Wyoming. Photo by A.J. Mangum.

When you make time a factor and equate the fastest time to winning, horsemanship tends to go out the window. Horses have no use for clocks; time limits are purely for the benefit of people.

- Peter Campbell

Peter Campbell and "Jackie," Lavelle Farms, Athens, Ohio. Photo by Nicole Lukins.

Chapter Eight
Cow Work

W***hen it comes to cow work*** and horsemanship, riders try to do too many things at once. They bring a cow out of the herd, then ride across the pen, stop the cow, turn the cow, drive the cow.

I tell students to break it down. Bring the cow out of the herd and give her a breather. Make a turn, give her a breather. That offers relief to the horse, too. It helps him get a grasp on the meaning of what's happening. Without that relief, the horse begins to wonder what the point of all this is, since he's getting no relief, no matter what he does.

It doesn't take much for a horse to make an effort at finding relief. Think about a fly landing on a horse. The horse gets bothered and swishes his tail to get rid of that fly. If you make the wrong thing a chore and the right thing easy, you will enter his mind like a key to a door. He will give you what you are asking for. Going forward, he'll begin to understand that trying for you is the path of least resistance.

One of the most common issues between horses and riders is that people don't give their horses enough relief. The reins aren't hooked to the horse's feet so they can't influence the movement of his hindquarters or front end, or play much of a role in getting him to sidepass or leg yield. Thus, when the rider goes to work a cow, the horse is pushy or confused, or doesn't respond to what the rider is asking.

When the horse is working a cow, it should either be in offense or defense mode -- whatever the rider needs at that

Peter Campbell and "Candy," Lavelle Farms, Athens, Ohio. Photo by Trina Campbell.

moment and time. Too often, the horse has been made so dull by the rider, it's hard to get the horse to understand what the rider's asking. Before cow work can be attempted with any hope of success, the rider has to work on horsemanship and know how to operate that horse's feet with the reins.

Riders need to learn stockmanship when it comes to effectively handling cattle. Try to read what's happening. Watch the cow. What do her ears do? What does her body do? What does she look like just before she turns left, or just before she runs off? Become aware of this expression and that feel will make it easier to handle stock.

If you simply charge into a herd, you lose the finer points necessary for this kind of work. You go in fast, the cow moves fast, and the horse's instinct for self-preservation kicks in. He figures that, at that pace, he'd better protect himself.

When I buckarooed, I worked cattle in open country, of

Peter Campbell and "Dixie," 5 Dot Land & Cattle Co., Benecia, California.
Photo by Trina Campbell.

course, and often worked alone. In that environment, you'd better know how to read cattle. If you apply too much pressure, they scatter. Most of the time, if you work slower, things work out better. And, when things work out better, the horse can actually contemplate the job at hand, take interest in it, and enjoy it. It might take some directing on the rider's part for the horse to understand what's being asked, but he'll soon get the picture and start to perform the job using his own mind, without the need for micromanagement by the rider.

Peter Campbell and "Molly," Loomis Ranch, Wheatland, Wyoming. Photo by Trina Campbell.

Peter Campbell aboard "Clara," Hat Creek Ranch, Wheatland, Wyoming.
Photo by Trina Campbell.

Special Thanks

I want to thank Tom and Margaret Dorrance, Ray and Carolyn Hunt, my mom, and all the folks out there along the trail that have helped me become who I am today.

"If we do not concern ourselves with who gets the credit, there is no end to what we can accomplish together."

I hope your journey through horsemanship is a journey about life.

Happy Trails,

Peter Campbell

Peter Campbell, Hat Creek Ranch, Wheatland, Wyoming. Photo by A.J. Mangum.

Peter Campbell is a lifelong horseman. He spends 40 weeks each year traveling from coast to coast, conducting clinics on colt starting, horsemanship, ranch roping and cow work. Peter and his wife, Trina, own the Hat Creek Ranch, a horse and cattle operation located outside Wheatland, Wyoming. Learn more about Peter at www.petercampbellhorsemanship.com.

Peter Campbell, "Molly" and "Sam the Dog," Hat Creek Ranch, Wheatland, Wyoming.
Photo by Trina Campbell.

CPSIA information can be obtained
at www.ICGtesting.com
Printed in the USA
LVIC090526011012
300390LV00001BB
* 9 7 8 0 9 8 5 3 4 2 5 2 4 *